Tucholsky Wagner Zola Scott Sydow Schlegel
Turgenev Fonatne Freud
Wallace
Twain Walther von der Vogelweide Fouqué Friedrich II. von Preußen
Weber Freiligrath Frey
Ernst
Fechner Fichte Weiße Rose von Fallersleben Kant Richthofen Frommel
Hölderlin
Engels Fielding Eichendorff Tacitus Dumas
Fehrs Faber Flaubert
Eliasberg Ebner Eschenbach
Feuerbach Maximilian I. von Habsburg Fock Eliot Zweig
Ewald Vergil
Goethe London
Mendelssohn Balzac Shakespeare Elisabeth von Österreich Dostojewski Ganghofer
Lichtenberg Rathenau Doyle Gjellerup
Trackl Stevenson Hambruch
Mommsen Thoma Tolstoi Lenz Hanrieder Droste-Hülshoff
Dach Verne von Arnim Hägele Hauff Humboldt
Reuter Rousseau Hagen Hauptmann Gautier
Karrillon Garschin
Damaschke Defoe Hebbel Baudelaire
Descartes Hegel Kussmaul Herder
Wolfram von Eschenbach Darwin Dickens Schopenhauer Rilke George
Bronner Melville Grimm Jerome Bebel
Campe Horváth Aristoteles Federer Proust
Bismarck Vigny Barlach Voltaire Herodot
Gengenbach Heine
Storm Casanova Tersteegen Gilm Grillparzer Georgy
Chamberlain Lessing Langbein Gryphius
Brentano Lafontaine
Claudius Schiller Kralik Iffland Sokrates
Strachwitz Bellamy Schilling
Katharina II. von Rußland Gerstäcker Raabe Gibbon Tschechow
Löns Hesse Hoffmann Gogol Wilde Gleim Vulpius
Luther Heym Hofmannsthal Klee Hölty Morgenstern
Roth Heyse Klopstock Kleist Goedicke
Luxemburg Puschkin Homer Mörike
La Roche Horaz Musil
Machiavelli Kierkegaard Kraft Kraus
Navarra Aurel Musset Moltke
Lamprecht Kind Kirchhoff Hugo
Nestroy Marie de France
Nietzsche Nansen Laotse Ipsen Liebknecht
Marx Ringelnatz
von Ossietzky Lassalle Gorki Klett Leibniz
May vom Stein Lawrence Irving
Petalozzi Knigge
Platon Pückler Michelangelo Kafka
Sachs Poe Liebermann Kock
de Sade Praetorius Mistral Zetkin Korolenko

The publishing house tredition has created the series **TREDITION CLASSICS**. It contains classical literature works from over two thousand years. Most of these titles have been out of print and off the bookstore shelves for decades.

The book series is intended to preserve the cultural legacy and to promote the timeless works of classical literature. As a reader of a **TREDITION CLASSICS** book, the reader supports the mission to save many of the amazing works of world literature from oblivion.

The symbol of **TREDITION CLASSICS** is Johannes Gutenberg (1400 – 1468), the inventor of movable type printing.

With the series, tredition intends to make thousands of international literature classics available in printed format again – worldwide.

All books are available at book retailers worldwide in paperback and in hardcover. For more information please visit: www.tredition.com

tredition was established in 2006 by Sandra Latusseck and Soenke Schulz. Based in Hamburg, Germany, tredition offers publishing solutions to authors and publishing houses, combined with worldwide distribution of printed and digital book content. tredition is uniquely positioned to enable authors and publishing houses to create books on their own terms and without conventional manufacturing risks.

For more information please visit: www.tredition.com

Bird Houses Boys Can Build

Albert Frederick Siepert

Imprint

This book is part of the TREDITION CLASSICS series.

Author: Albert Frederick Siepert
Cover design: toepferschumann, Berlin (Germany)

Publisher: tradition GmbH, Hamburg (Germany)
ISBN: 978-3-8491-8639-5

www.tredition.com
www.tredition.de

Copyright:
The content of this book is sourced from the public domain.

The intention of the TREDITION CLASSICS series is to make world literature in the public domain available in printed format. Literary enthusiasts and organizations worldwide have scanned and digitally edited the original texts. tradition has subsequently formatted and redesigned the content into a modern reading layout. Therefore, we cannot guarantee the exact reproduction of the original format of a particular historic edition. Please also note that no modifications have been made to the spelling, therefore it may differ from the orthography used today.

BIRD HOUSES
BOYS CAN BUILD

SIEPERT

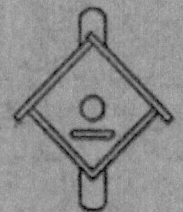

FOREWORD.

Years ago a country boy heard or read that if a simple box having a hole of a certain size were set upon a post in March or early April it would not be long before bluebirds would be around to see if the place would do as a summer cottage. So he took an old paint keg such as white lead is sold in, nailed a cover across the top, cut an opening in the side and then placed it on a post ten or twelve feet high. Only a day or two passed before a soft call-note was heard, a flash of blue, and the songster had arrived. His mate came a few days later and the paint keg with its tenants became the center of interest in my life. A second brood was reared in midsummer and when the cool days of September came a fine flock left for the South. Each year the house was occupied until the post decayed and the paint keg fell down, but in memory the sad call-note is still heard when spring comes, for it is house hunting time once more, and the bluebirds are looking for the home they had known.

That boys elsewhere may know the joy of the companionship of birds, this little book is written. Birds will come and live near the houses of men whenever food and water are to be had, safety from enemies is given, and when homes are built for them to replace the shelters nature offered before men came with their cultivated fields and crowded cities. The following pages give pictures and drawings of houses that boys have built and in which birds have lived. These houses are planned for the species of birds that have become accustomed to civilization so that they will inhabit the houses put up for them.

The author is indebted to Professor Chas. A. Bennett of Bradley Institute and Mr. L. L. Simpson of The Manual Arts Press for helpful suggestions and encouragement; to John Friese for making the drawings; and to the following for the use of the originals of the illustrations which tell most of the story.

Edward G. Anderson, Seattle, Wash. Figs. 32, 33, 34, 36, 39, 54, 55, 56, 57.

[Pg 4]

Frank H. Ball, Pittsburgh, Pa. Figs. 12, 29, 45, 66, 67.

Leon H. Baxter, St. Johnsbury, Vt. Figs. 21, 22.

F. D. Crawshaw, University of Wisconsin, Madison, Wis. Figs. 11, 40, 41, 42, 43, 44.

Donald V. Ferguson, St. Paul, Minn. Figs. 9, 28, 38, 62.

Geo. G. Grimm, Baltimore, Md. Fig. 14.

C. M. Hunt, Milton, Mass. Figs. 46, 52.

H. A. Hutchins, Cleveland, O. Figs. 15, 16, 17, 18, 19.

Elmer Knutson, St. Cloud, Minn. Figs. 30, 31.

National Association of Audubon Societies, 1974 Broadway, New York City. Figs. 1, 2, 3, 4, 5, 6, 7, 8.

Chas. Tesch, Milwaukee, Wis. Fig. 64.

The Crescent Co., Toms River, N. J. Figs. 35, 49, 50.

United States Department of Agriculture Bulletins; Figs. 20, 51, 65.

Youths Companion, Perry Mason Co., Boston, Mass. Figs. 58, 59, 60, 61.

Albert F. Siepert.

Peoria, Ill., March, 1916.

CONTENTS

Birds That Live in Nesting Boxes

Bluebird — robin — chickadee — wren — house finch — woodpecker — flicker — martin

Construction of Bird Houses

Dimensions of nesting boxes — houses of sawed lumber — rustic houses — cement and stucco houses

Placing Houses

Feeding Shelves and Shelters

Foods

Bird Baths

Bird Enemies

Men — ants and vermin — English or house sparrow — sparrow traps

Bird House Exhibitions

Bibliography

Bird and bird house literature

BIRDS THAT LIVE IN NESTING BOXES.

Certain varieties of birds will nest in homes built for them if these houses are of the right shape and dimensions. Other birds may be just as desirable but do not build nests and rear their young in boy-made nesting boxes. We are therefore mainly concerned with the first group which select cavities in trees for their homes if nothing better is to be found.

FIG. 1. BLUE BIRDS, ADULTS AND YOUNG BIRD.

BLUEBIRD.

This bird may be found during the summer months in most of the states east of the Rocky Mountains, Figs. 1 and 59. It spends its winter in the southern states and southward, returning north in March and April. The principal items of food are grasshoppers, caterpillars and beetles. It should have a house measuring about 5" in length and width, inside measurements, and 8" or more in depth. The entrance hole should be 1-1/2" in [Pg 8] diameter and placed near the top, so that the young birds cannot get out until strong enough to have some chance of escape from their enemies after they leave the nest. While authorities differ as to the need of cleaning after a season's use, it seems wise to provide the house with some device whereby the bottom may be removed for such purposes. Houses for this species are shown in Figs. 11, 21, 22 and 24.

FIG. 2.

ROBIN.

Robins usually announce the coming of spring when they return to their breeding grounds in the northern states, where they are general favorites. Figs. 2 and 60. The nest is usually built of mud and lined with grasses; placed in the fork of a tree or on some sheltered ledge. Robins take kindly to nesting shelves [Pg 9] put up for them and it is well to put up several since but one brood is reared in each nest built. This old nest should be removed after the young birds have gone. A simple shelf is shown in the lower left hand corner of the photograph, Fig. 24, as well as in Figs. 20 and 49.

FIG. 3.

CHICKADEE.

The chickadee is one of the brave little spirits who spends the entire winter with us, Fig. 3. We can be of considerable service to him during the cold weather by providing food shelters. During the summer months his home is usually found in [Pg 10] some decaying stump, hence nesting boxes of the rustic type placed in some remote spot of the orchard or park are most attractive to him.

WREN.

When all other song birds fail to take advantage of a house built for them, the wren may still be counted on. Almost any sort of home from a tin can or hollow gourd on up is satisfactory if put in a safe place and provided with an opening 1" or slightly less in diameter, so the sparrows must stay out, Figs. 4 and 5. Good homes are shown in Figs. 10, 14, 15, 16 and others.

FIG. 4. WREN AND RUSTIC HOUSE.

HOUSE FINCH.

The house finch has made many enemies because of its fondness for cultivated fruits and berries. However, it has some [Pg 11] redeeming features in its song and beauty. The nest is usually placed in the fork of a limb—evergreens being favorite nesting places. The house shown in Fig. 51 is suitable for these birds but is also acceptable to wrens.

FIG. 5. WRENS.

FIG. 6. FLICKER.

WOODPECKER.

The favorite of this interesting family is the little downy, Fig. 7. Living largely upon harmful grubs and insects, this bird does an immense amount of good by protecting our forests from insect scourges. Woodpeckers do not build nests as most birds do, but excavate a deep cavity in some dead tree leaving a quantity of chips at the bottom on which the eggs are laid. Nesting boxes should be of the rustic type made as shown in Fig. 12, leaving some sawdust mixed with a little earth in the cavity. These houses should be placed on trees in a park or orchard. Boys should be able to tell the difference between the woodpeckers [Pg 12] beneficial to man and the sapsucker whose misdeeds often cause considerable damage to fruit trees. A nuthatch is also seen in Fig. 7 enjoying a meal of sunflower seed.

FIG. 7. DOWNY WOODPECKER (ABOVE) NUTHATCH (BELOW).

FLICKER.

The flickers spend much of their time on the ground in search of ants which form the larger percentage of their food. Since ants sometimes cause considerable trouble for other birds, a pair of flickers are worth cultivating for the sake of the work [Pg 13] they can do. Artificial nesting boxes of sufficient depth and size are quite readily used, Figs. 6, 20 and 25.

MARTIN.

Nearly everyone knows swallows of one variety or another. The most beautiful of the family are the martins, Fig. 8. This bird is of great service against the inroads of wasps, bugs and beetles. It prefers to live in colonies even though the males fight bitterly at times. Martin houses should have at least several rooms, each separate from all the others. Houses have been built to accommodate fifty and more families. Smaller ones are shown in Figs. 8, 9, 13 and 45.

FIG. 8. A MARTIN COLONY.

[Pg 14]

FIG. 9. THE PEER GYNT COTTAGE FOR MARTINS.

Fig. 9 is a miniature reproduction of Peer Gynt's cottage for a martin house. This house was not only an attractive thing to make, but martins selected it for their home during the past summer.

[Pg 15]

CONSTRUCTION OF BIRD HOUSES.

Bird houses may be divided into three main classes: (1) those made of sawed lumber to specified dimensions; (2) the rustic type made of (a) slabs of wood with the bark left on, or (b) pieces of tree trunk, or (c) of sawed lumber trimmed with bark or twigs; and (3) cement or stucco houses. In each case the entrance should slant slightly upward to keep the rain out.

FIG. 10. WREN HOUSES.

Almost any sort of lumber may be used, but birds take most readily to that which has been weathered out of doors. A kind should be used which does not warp or check badly; white pine and cypress meet these requirements and are worked with ease. Yellow poplar is used and cedar with or without the bark left on has its friends for houses of the first or second classes.

Nesting boxes of sawed lumber should be painted on the outside to improve their appearance and to preserve them against the effect of the weather. It is often wise to leave a small amount of unpainted surface around the entrance, and all paint should be thoroughly dry before houses are expected to be occupied. Colors selected will depend somewhat upon the neighborhood, but white, grey, dull greens or browns are often used.

[Pg 16]

DIMENSIONS OF NESTING BOXES.

The following table, copied from Farmers Bulletin, No. 609, U. S. Dept. of Agriculture, gives in small space valuable information about dimensions that experience and investigation have indicated as good for particular varieties of birds. This list includes many varieties that do not commonly live in houses built for them, however. As time goes on, we may expect to find more of these birds living in our nesting boxes because they are apt to seek the same sort of home as the one in which they were reared. The table is given to be of service to those wishing to plan new houses not shown here.

Dimensions of nesting boxes for various species of birds.

Species.	Floor of cavity.	Depth of cavity.	Entrance above floor.	Diameter of entrance.	Height above ground.
	Inches.	Inches.	Inches.	Inches.	Feet.
Bluebird	5 by 5	8	6	1-1/2	5 to 10
Robin	6 by 8	8	[1]	[1]	6 to 15
Chickadee	4 by 4	8 to 10	8	1-1/8	6 to 15
Tufted titmouse	4 by 4	8 to 10	8	1-1/4	6 to 15
White-breasted nuthatch	4 by 4	8 to 10	8	1-1/4	12 to 20
House wren	4 by 4	6 to 8	1 to 6	7/8	6 to 10
Bewick wren	4 by 4	6 to 8	1 to 6	1	6 to 10

Carolina wren	4 by 4	6 to 8	1 to 6	1-1/8	6 to 10
Dipper	6 by 6	6	1	3	1 to 3
Violet-green swallow	5 by 5	6	1 to 6	1-1/2	10 to 15
Tree swallow	5 by 5	6	1 to 6	1-1/2	10 to 15
Barn swallow	6 by 6	6	[1]	[1]	8 to 12
Martin	6 by 6	6	1	2-1/2	15 to 20
Song sparrow	6 by 6	6	[2]	[2]	1 to 3
House finch	6 by 6	6	4	2	8 to 12
Phoebe	6 by 6	6	[1]	[1]	8 to 12
Crested flycatcher	6 by 6	8 to 10	8	2	8 to 20
Flicker	7 by 7	16 to 18	16	2-1/2	6 to 20
Red-headed woodpecker	6 by 6	12 to 15	12	2	12 to 20
Golden-fronted woodpecker	6 by 6	12 to 15	12	2	12 to 20
Hairy woodpecker	6 by 6	12 to 15	12	1-1/2	12 to 20
Downy woodpecker	4 by 4	8 to 10	8	1-1/4	6 to 20
Screech owl	8 by 8	12 to 15	12	3	10 to 30
Sparrow hawk	8 by 8	12 to 15	12	3	10 to 30

Saw-whet owl	6 by 6	10 to 12		10	2-1/2	12 to 20
Barn owl	10 by 18	15 to 18		4	6	12 to 18
Wood duck	10 by 18	10 to 15		3	6	4 to 20

[1] One or more sides open.

[2] All sides open.

HOUSES OF SAWED LUMBER.

The boy with an outfit of tools at home, or with a teacher of manual training interested in birds, can make all of the houses to be described in this section. Figs. 10 and 11 show simple houses for wrens and bluebirds. Drawings for this type of house [Pg 17] are shown in Figs. 14, 15 and 21. While the surfaces of lumber used for these houses may or may not be planed, care must be taken that all pieces are sawed or planed to the correct sizes with edges and ends square and true so there will be no bad cracks for drafts and rain to enter. Be careful to nail the pieces together so that they will not have occasion to crack or warp. A good way to save time and lumber is to prepare a piece of stock, getting it of the right thickness, width and length, and then to saw up this stock on lines carefully laid out as shown in the drawings of the bluebird and wren houses, flicker nest, robin shelf and finch house. The most difficult houses to build are those for martins. In Fig. 45 is given a drawing for a small home arranged to care for eight families, while the photographs, Figs. 8, 9, 38, 66 and 67 show larger, finer and more difficult houses. The [Pg 18] doors or openings are 2-1/2" in diameter and can be made with an expansion bit or a key-hole saw. All of these houses are to be made so they may be cleaned. Sometimes the bottom is hinged on two screws or nails, and held in proper place by a dowel (bluebird house, Fig. 21); or screwed in place (wren house, Fig. 21, and martin house, Fig. 45); or hinged and held in place by a brass spring (wren house, Fig. 14).

FIG. 11. HOUSES FOR WRENS AND BLUEBIRDS.

FIG. 12. RUSTIC HOUSES.

RUSTIC HOUSES.

The first group of houses of this type are shown in Figs. 12, 35 and 36. These are made of slabs of wood with the bark left on, and in some cases, of the bark alone if it can be secured of sufficient thickness. It is usually a good plan to drive a sufficient number of nails into the bark to keep it in place, otherwise it will drop off. Houses such as these attract birds that would avoid a freshly painted imitation of some large residence or public building. Figs. 20 and 37 show houses made of a section of a tree split or sawed in halves, the nest cavity hollowed out, and then fastened together again with screws. The top should be [Pg 19] covered with a board or piece of tin to keep out rain. The third division of this type of house is made of sawed lumber and then trimmed with bark or twigs. In this way the same frames may be made to appear as very different bird houses when completed. Such houses are shown in Figs. 30 to 34. Sometimes a pail is used for the frame and then covered with bark, as the center house of Fig. 28. This house has a partition placed halfway up making it a two family apartment, and is provided with ventilating and cleaning devices.

CEMENT AND STUCCO HOUSES.

Houses may be cast of concrete as Fig. 39. This requires a mold or form, and takes considerable planning to insure success. A form is made whose inside dimensions are those of the outside of the bird house, and of the desired shape. A second form, or core, to be placed inside of the first form, is made as large as the inside of the bird house. The two forms must be mounted so they will remain in the right relation while concrete is placed in the space between them. After this has set, the forms may be removed, cleaned and used again. The roof is generally made separately and put in position last. Or the roof can be cast as a part of the house in which case the bottom is inserted last. Birds do not take as kindly to this type of house as a rule, as to those made of wood.

FIG. 13. STUCCO HOUSE FOR MARTINS.

The stucco house has many possibilities. Fig. 38, shows a group of such houses designed to match the general appearance of garages in good residence districts. The frame is made of wood and t Pg020 co applied by one of the methods in use on large houses. Seventh grade boys have made such houses, using 3/8" material for the frames, tacking on wire netting and then plastering each side of the house in turn with concrete. The sides were given a pebble-dash surface, while the roof was finished with a steel trowel to give a smooth surface that will shed water readily, Fig. 13.

[Pg 20]

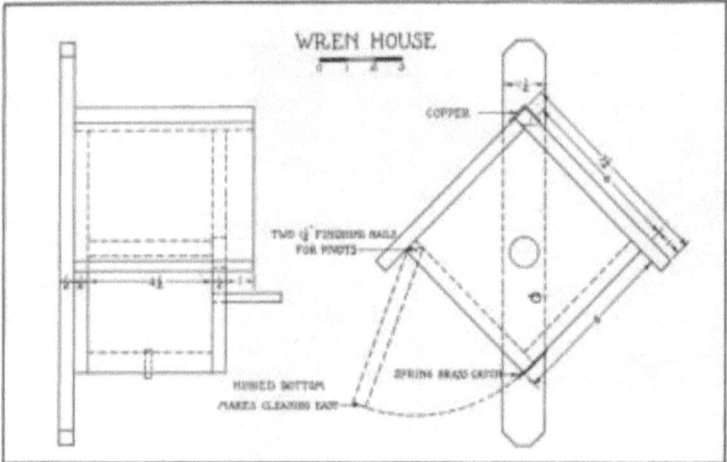

FIG. 14.

[Pg 21]

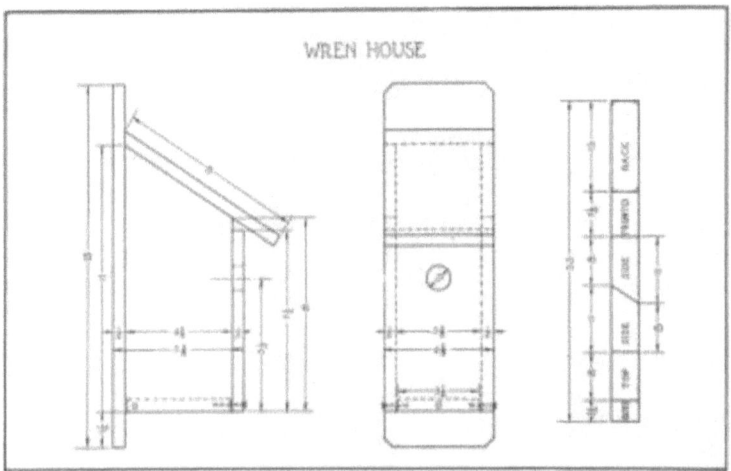

FIG. 15.

[Pg 22]

A CAREFULLY PLANNED PROJECT IS SHOWN IN VARIOUS STAGES OF COMPLETION IN FIGS. 15-19.

FIG. 16. WREN HOUSE.

FIG. 17. ECONOMY OF TIME AND MATERIAL WHEN LAID OUT IN THIS MANNER.

[Pg 23]

FIG. 18. ASSEMBLING BIRD HOUSES.

FIG. 19. FINISHING BIRD HOUSES.

[Pg 24]

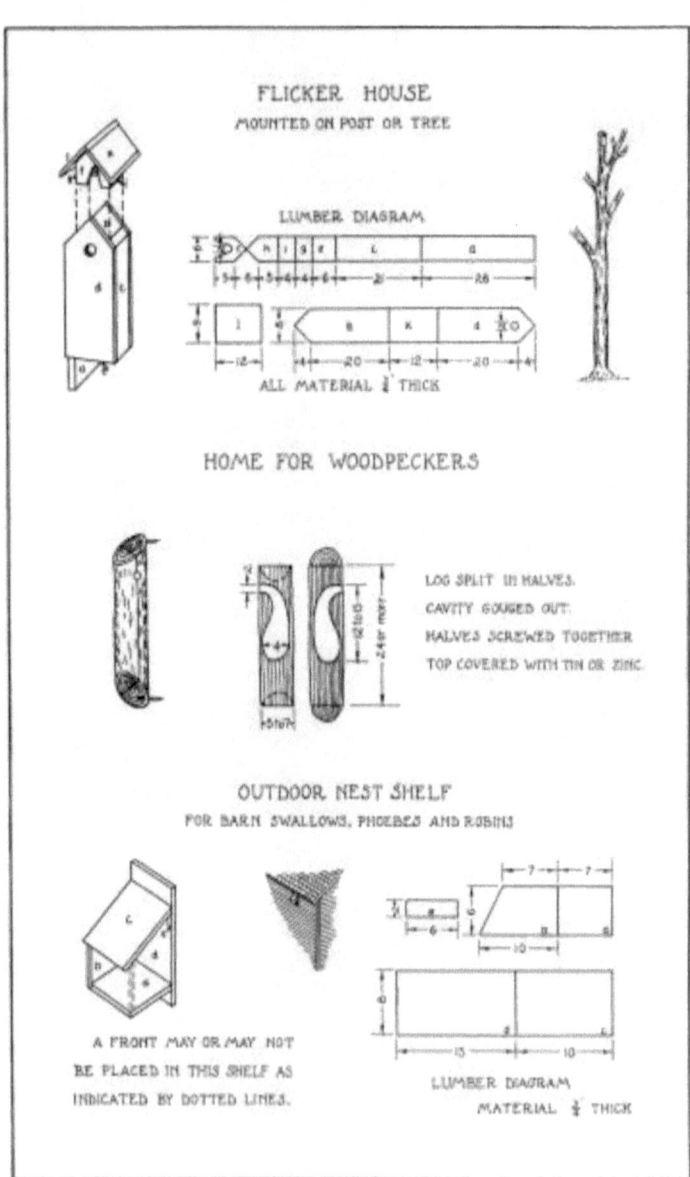

FIG. 20
(Click image for enlarged view.)

[Pg 25]

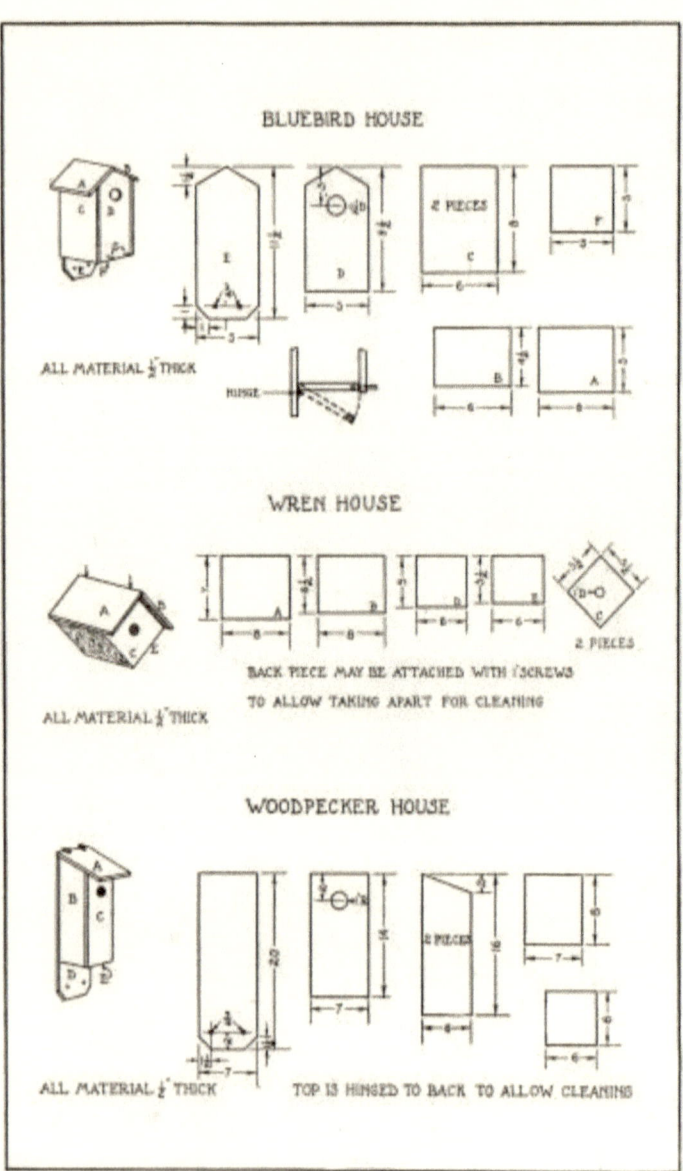

FIG. 21.
(Click image for enlarged view.)

[Pg 26]

FIG. 22. HOUSES BUILT BY STUDENTS AT ST. JOHNSBURY, VT.

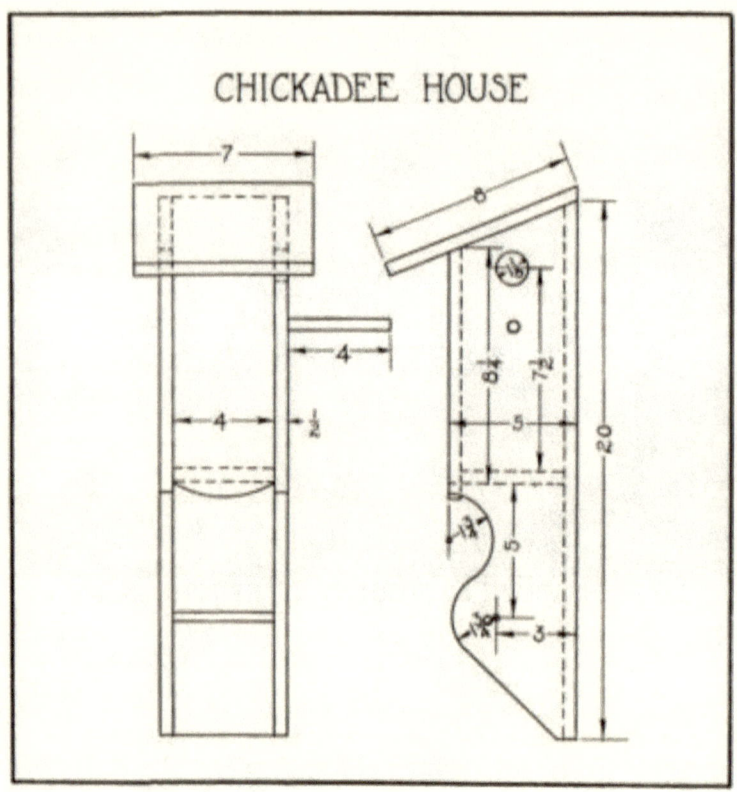

FIG. 23.

[Pg 27]

FIG. 24. WREN, BLUE BIRD AND ROBIN HOUSES.

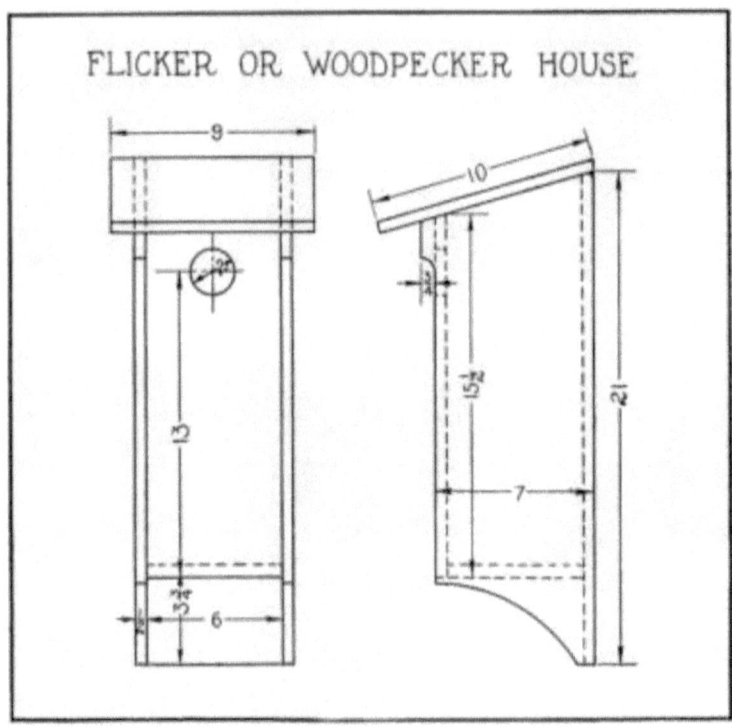

FIG. 25.

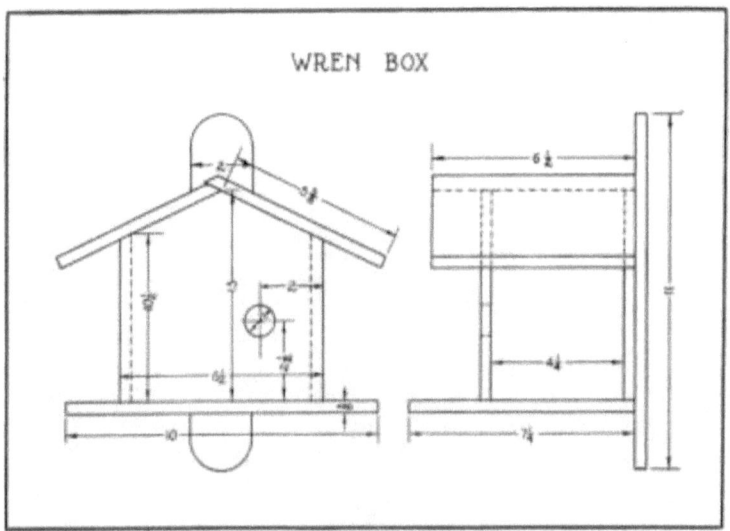

FIG. 26.

[Pg 29]

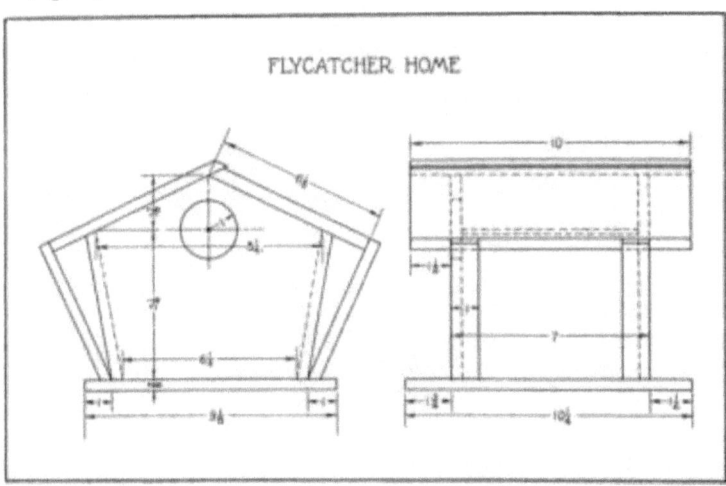

FIG. 27.

FIG. 28. RUSTIC HOUSES MADE BY ST. PAUL, MINN. BOYS.

FIG. 29. RUSTIC HOUSES MADE BY PITTSBURGH, PA. BOYS.

FIG. 30. SIMPLE LOG AND BIRCHBARK CONSTRUCTION, HOUSES FOR WRENS, BLUEBIRDS, ETC.

FIG. 31. BIRCH BARK HOUSES.

FIG. 32. GOOD TYPES OF SMALL HOUSES.

FIG. 33. GOOD TYPES OF SMALL HOUSES.

FIG. 34. A QUAINT BIRD HOME.

FIG. 35. A HOUSE OF CEDAR SLABS FOR JENNY WREN.

FIG. 36. MAKING BIRD HOUSES TO ORDER.

[Pg 34]

FIG. 37. CONSTRUCTION OF A WOODPECKER'S HOME. A MARTIN, OR TREE SWALLOW HOME.

FIG. 38. STUCCO HOUSES.

[Pg 35]

FIG. 39. CONCRETE HOUSES.

FIG. 40. READY TO PLACE FINISHED HOUSES.

[Pg 36]

PLACING HOUSES.

The table given on page 16 states the height from the ground that different species of birds seem to prefer for their nests, to which several suggestions may be added. The houses should be so located that cats and other bird enemies do not have easy access to them. The openings ought to be turned away from the directions from which storms and winds most often come; and the house must hang or tilt so rain does not run in at the entrance. Such birds as the woodpeckers spend most of their time [Pg 37] in the trees and so do not take as readily to a house set on a pole out in the open as martins or bluebirds. Flickers are seen on the ground a good share of the time in search of their favorite food, and so will frequently live in houses nailed to fence posts. Houses are more apt to be occupied if placed in position in fall or winter before the spring migration, especially houses made of freshly dressed or newly painted wood. However, such birds as the robin and bluebird rear more than one brood each season and so a house set up in May or June may have a tenant. Figs. 40 to 44 show boys of the University of Wisconsin High School placing some of the houses they had made.

FIG. 41.

FIG. 42.

FIG. 43.

FIG. 44.

FEEDING SHELVES AND SHELTERS.

Nesting boxes make their appeal to but a part of the birds of any community. These attract during the early spring and summer months. Many other species are worth having in our orchards and gardens for their songs and their activity in destroying insects and weed seeds. To these some other attraction [Pg 41] than nesting boxes must be offered. Then again, many birds would spend a longer time with us if a certain food supply were assured them. A simple suet feeder is shown in Fig. 45. The birds cling to the chicken wire while eating. A feeding box for seed-eating birds is given in Fig. 46. Fig. 47 gives a shelf to be nailed to the sunny side of a building, while Fig. 48 shows a somewhat similar type to be fastened to a window sill, making it possible to observe the birds that come to dine. Birds that hesitate to come close to buildings may be attracted by the feeders set out in the open. Fig. 50 shows a feeder mounted on an iron pipe so it can be turned in any direction. This feeder has one end closed by a pane of glass, and is to be turned so that prevailing winds do not enter. Fig. 49 shows a feeding shelf for winter use which makes an acceptable robin nesting shelf in spring. In Fig. 53 is given a feeder mounted on a base with a vane so the adjustment takes place automatically. Figs. 51 and 52 show two food shelters considerably more difficult to construct. They have glass on all sides, and are open at the bottom so that birds can enter or leave at will. Fig. 30 shows a simple food shelter offering some protection against rain and snow, while a very attractive group of shelters are given in Figs. 54, 55, 56 and 57. If you look closely you may see "Mabel" in the right hand feeder in Fig. 54. The builder of these shelters found her so positive about her rights—since she discovered the food supply—that he has been obliged to put up the others to keep peace.

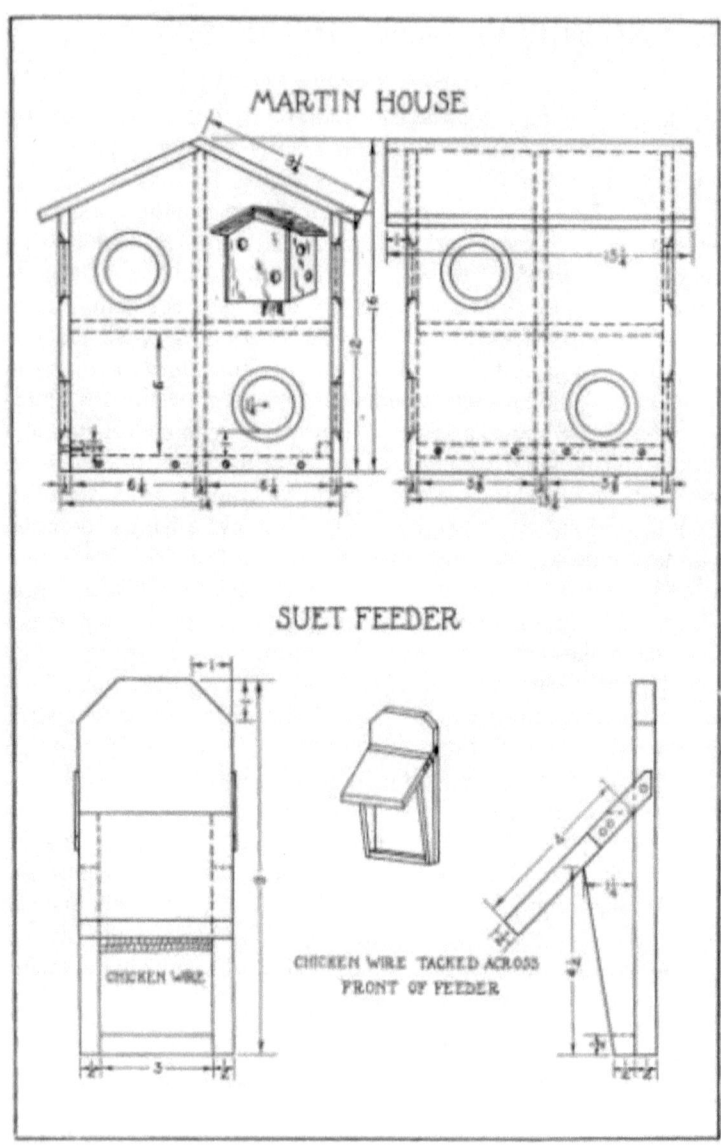

FIG. 45.
(Click image for enlarged view.)

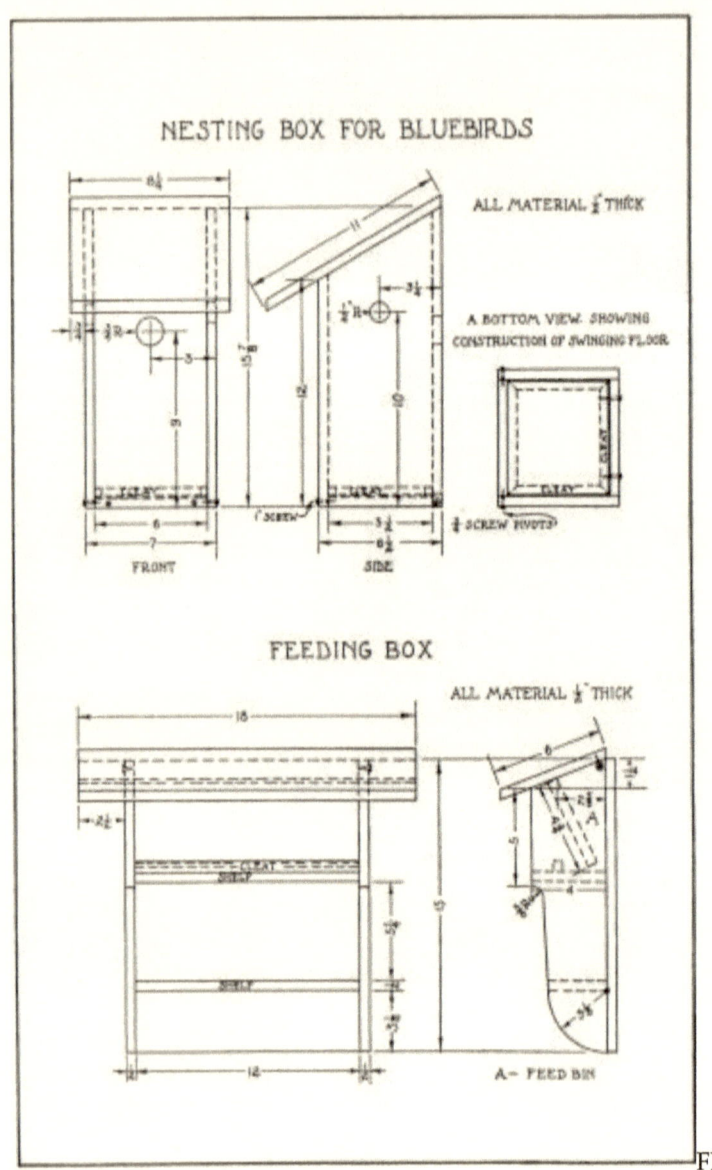
FIG. 46.

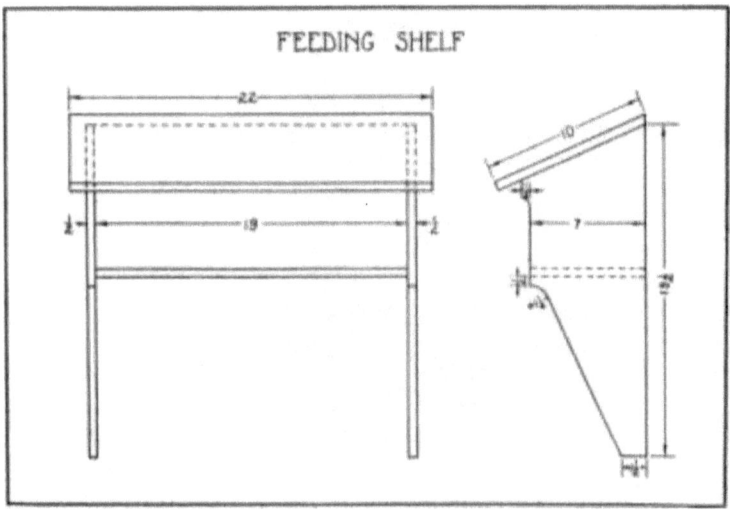

FIG. 47.

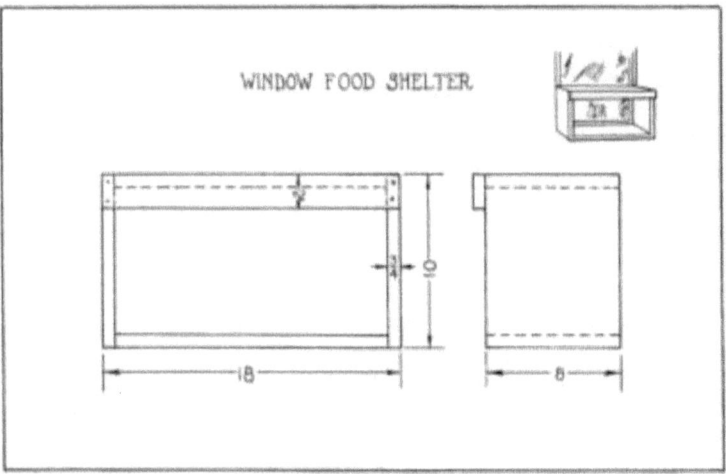

FIG. 48.

FIG. 49. ROBIN SHELF OR FEEDING SHELF.

FIG. 50. HILBERSDORFER FOOD HOUSE.

[Pg 42]

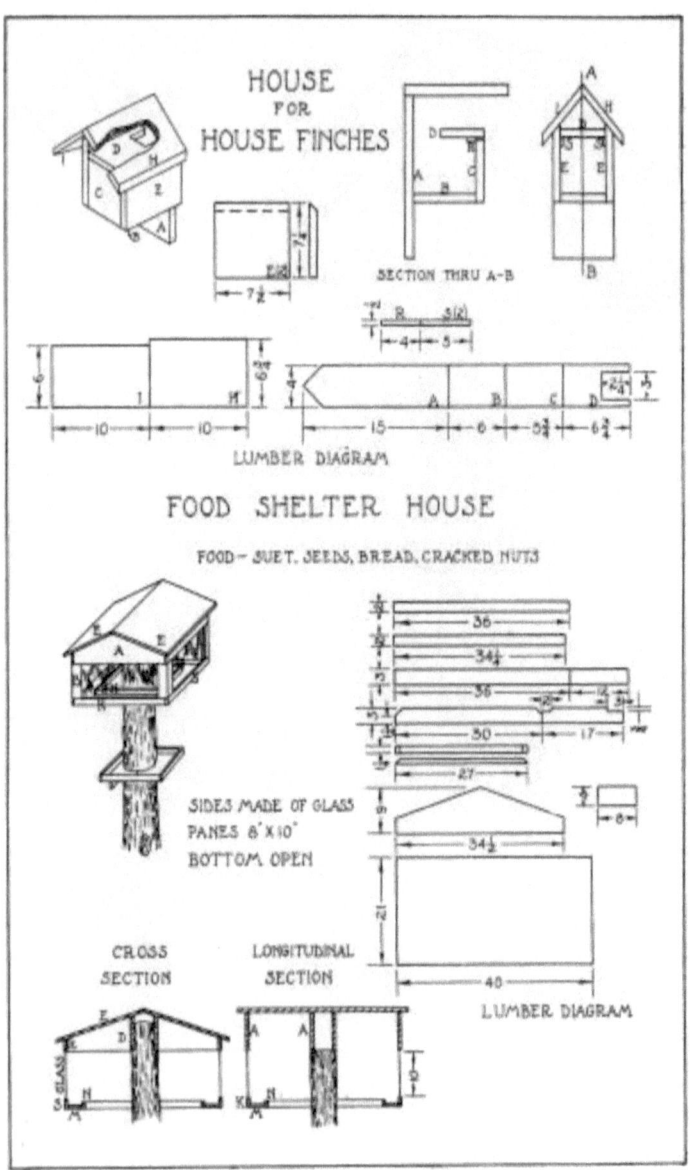

FIG. 51.

[Pg 43]

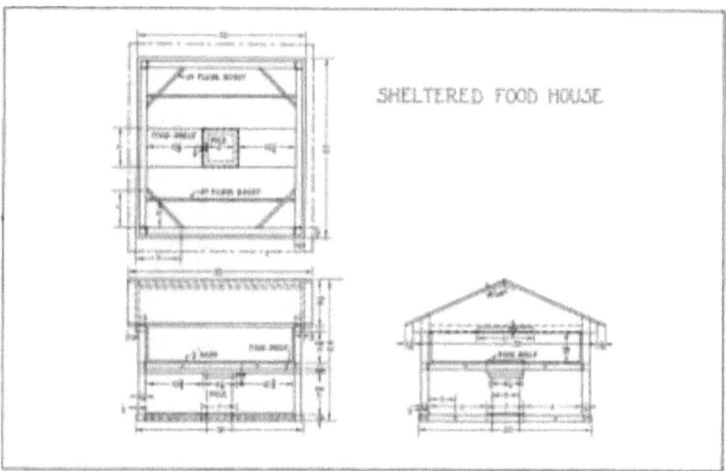

FIG. 52.
(Click image for enlarged view.)

[Pg 44]

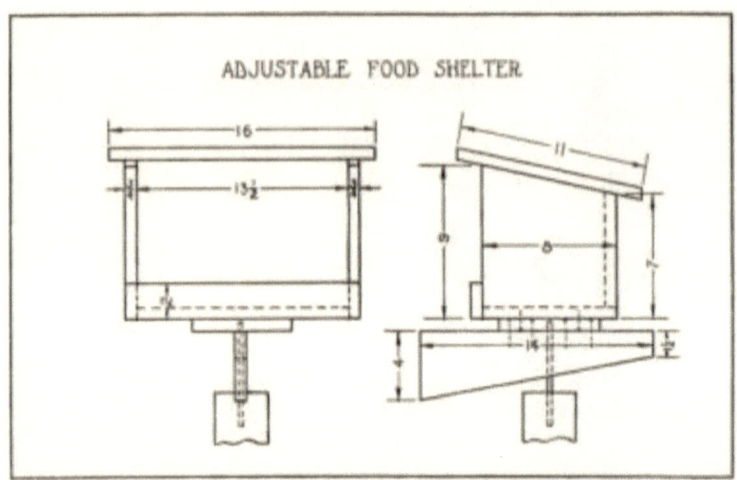

FIG. 53.

FIG. 54.

[Pg 45]

FIG. 55. WHEN THE SHELTER IS MOST NEEDED.

FIG. 56.
BIRDS SHOULD FIND THESE FOOD SHELTERS BEFORE BAD WEATHER COMES.

FIG. 57. A "JAPANESE" EFFECT.

The window-sill lunch counter shown in Figs. 58 and 61 is a most effective way to study birds at close range. The window selected for this purpose should be on a quiet and sheltered side of the house if possible. If trees and shrubbery are near at hand birds are more likely to be attracted. Branches of thorn apples, alders and evergreens are fastened firmly to the window frames to dress the lunch counter on the outside while house plants or [Pg 46] at least a curtain should be placed on the inside as a screen. Fig. 59 shows how particular varieties of birds may be attracted by offering favorite foods while Fig. 60 gives an idea of what kindness will do.

FIG. 58. THE BIRD WINDOW SEEN FROM INSIDE THE ROOM.

FIG. 59. BLUEBIRDS ATTRACTED TO THE WINDOW SILL BY MEAL WORMS.

FIG. 60. A ROBIN ABOUT TO EAT FROM THE CHILDREN'S HANDS.

FIG. 61. THE WINDOW-SILL LUNCH COUNTER FROM OUT-SIDE.

FOODS.

Food shelters become centers of interest in proportion to the number of birds attracted to them. The kind of food placed there determines in time the kind of birds that will be found frequenting them. Seed-eating birds are readily attracted by the use of small grains such as oats and wheat. However, every [Pg 47] farmer finds a quantity of weed seeds upon cleaning his seed grain, which proves very acceptable to chickadees and blue jays. Bread crusts or crumbs, crackers and doughnuts may be placed in the food shelter with the knowledge that the birds will eat them. For those of the city who would need to buy seeds, it will be just as well to get hemp, millet, canary seed and sunflower [Pg 48] seed, together with the small grains and cracked corn for foods. Suet, scraps of meat and various vegetable scraps, such as celery, lettuce, apples, raisins, and the berries of various bushes, if they can be obtained, are relished. Bluebirds seem fond of meal worms such as develop in old cereals. All birds require water and frequently suffer because this is not to be had. If it is possible to meet this need a great service is rendered. Finally, when the ground is snow covered, many birds appreciate a supply of sand and finely ground poultry grit. Many birds are lost each winter because of insufficient food during inclement weather, that if cared for would remain near neighbors in the summer to wage war upon insect pests.

BIRD BATHS.

The best bird baths have to meet two requirements: (a) clean, fresh water, and (b) safety from enemies. Almost any shallow dish will meet both requirements if properly placed and cared for.

FIG. 62. THE PALM GARDEN EXHIBIT OF BIRD BATHS, ETC., ST. PAUL, MINN.

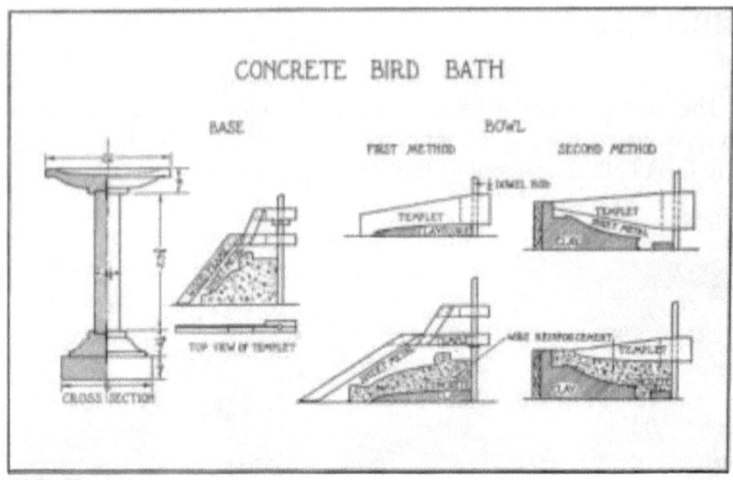

FIG. 63.

Fig. 62 shows several baths made of concrete. The pedestal and basin are made of two separate pieces, and are cast in a form or mold. A more difficult concrete bath is shown in Fig. 63. This project is made in four pieces. The base consists of two parts, the bottom being cast in a form made of 1/2" or 7/8" stock. The upper part is "swept up" by means of the templet shown, which revolves about an iron rod or a dowel-rod firmly fastened above, and held below in a hole bored in a temporary base of wood. The column is cast in a mold made of sheet tin or galvanized iron run thru tinners' rolls, and held by means of several wires twisted about it. When this is being cast two pieces of iron rod are inserted as shown which are to pass into both bowl and base to make the whole job firm. The bowl may be swept by either of two methods. The first consists of the making of two templets. With the first templet a core of clay is swept up of the desired depth and diameter. Then concrete is placed over this core, which has previously been treated to a coat of oil. Woven wire is cut into a circular shape and bent to approximate the curve of the bowl. More concrete is placed over this, and swept up by means of the second templet. Some difficulty will be experienced in removing the templet if undercut as much as shown; however, [Pg 51] the mark where it was taken off can easily be troweled smooth again. The finished pieces are now assembled with a small quantity of "neat" cement in each joint. The second method for making the bowl begins with the making of templets cut on the opposite side of the outline, as compared with the first method. A box is then nailed up and a clay or plaster-of-paris base made. This is oiled, and the concrete put in place. In this case a wetter mixture than in the first case should be used. The second templet is then used to strike off the inside of the bowl. After this has set the pieces may be assembled as before.

BIRD ENEMIES.

One sometimes wonders that birds manage to exist and to actually increase in numbers. Possibly the first group of enemies should include men and boys who kill adult birds, leaving the fledglings to starve, or who rob the nests of eggs. It is the writer's belief that every boy who makes one or more of the projects in this booklet, and sees it occupied, will become one of a growing number who will care for instead of destroy the birds of his neighborhood. Further, if every man who now thoughtlessly or willfully destroys birds, knew the real money value of the work birds do, he would build or buy houses and food shelters to increase Nature's best friends to mankind.

The second group of enemies include ants and other vermin which at times infest nests and nesting boxes, snakes, squirrels, mice and rats. Protection against this group is afforded by bands of tin about the pole, or spikes of wood or metal pointing downward so that access is impossible by climbing up the pole. Another protection is to make the entrance holes small enough to admit only the occupant for whom the house was intended. Of course, the houses for the larger birds must be protected in other ways. Charles Tesch of Milwaukee suggests a sticky fly paper compound made of resin (melted) and castor oil as a preventive for the inroads of the small red ant, if suitable support is available.

The final group consists of the two worst foes of bird life, cats and English or house sparrows. If you really value the birds that have been reared in the house you have built you may need to get up *early* more than one morning when the youngsters [Pg 52] leave the nest to protect them from the highly respectable (?) tabby that lives possibly next door if not at your own house. It often comes to a choice between cats and birds: and the cats may be disposed of in two ways—the right kind of box traps for the homeless and unknown robbers, and an air rifle with sufficient "sting" for the trespasser from next door. A few lessons of this kind usually have some effect.

The English or house sparrow was introduced into this country about a half-century ago. It has spread over practically all of the

United States and Southern Canada. Possibly no bird has exhibited such powers of adapting itself to new conditions. The sparrow is no respecter of places for locating its nest. It lives on a variety of foods changing from one to another as the necessity arises. In spite of opposition, this bird is constantly on the increase, so much so that in many cases more desirable native birds have been obliged to leave. The sparrow is filthy and quarrelsome, and lives mainly upon valuable small grains in every case where this is possible. There are two methods possible which afford partial relief: (1) traps and (2) driving them away with an air-rifle. Traps are usually successful for a comparatively brief time, since the sparrows soon associate the trap with danger and so avoid it. A very successful type of woven wire trap is advocated by the Department of Agriculture but is probably beyond the ability of the average boy to make well. It sells by commercial manufacturers of bird supplies for about $4.00. This trap works all the year around as it depends upon the attraction of food. Fig. 65 gives a simple, yet effective trap. However, it requires the presence of some hidden observer to spring it at the right moment. Another type of trap is based upon the nest-house idea. Its effectiveness is limited largely to the nesting season, though it may be used by the birds for shelter. One of the most efficient traps was invented by Charles Tesch, of Milwaukee, Wis. Its principle is that of a tipping chamber leading into a sack thru a chute. Fig. 64 gives the dimensions to be followed in making such a trap. The inventor says that the bag should be far enough away from the box to make certain that the victim has no chance to tell the others what happened to him by chirping, otherwise they will no longer enter the trap. The box must be perfectly tight in order to prevent drafts from issuing thru the entrance which will cause sparrows to keep away. If a few feathers are glued or shellacked to the tipping chamber floor, the sparrow is often attracted more strongly. The bag should be examined frequently to liberate bluebirds and wrens, who may have been caught.

[Pg 53]

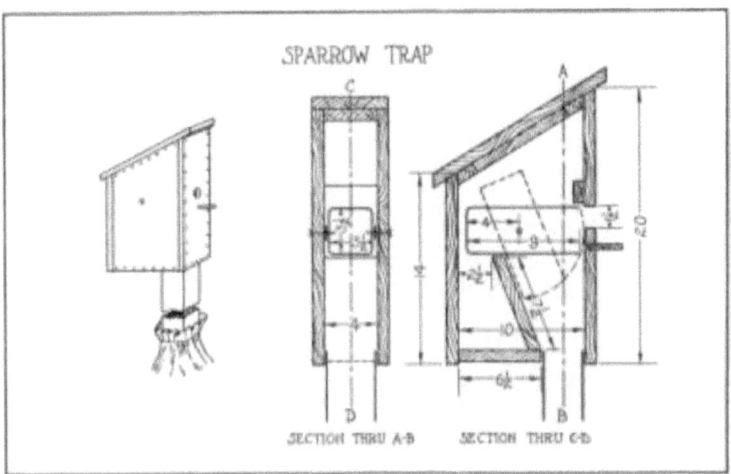

FIG. 64.

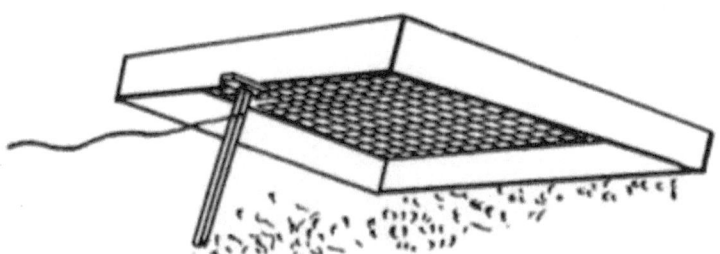

FIG. 65.

[Pg 54]

However, fighting bird enemies without the cooperation of neighbors is not an easy matter. In the case of sparrows, so many more are left that traps alone are ineffective. An airgun properly used offers some help in the city to drive them away from the premises, while a shot gun or 22 caliber rifle are more effective in the country. If every sparrow nest were torn down and no place given them in your neighborhood, the pest is likely to avoid your grounds. Finally, keep nesting boxes free from sparrows while the owners are away for the winter.

BIRD HOUSE EXHIBITIONS.

Many cities are beginning to do excellent work along the lines of bird preservation and attraction. This usually leads into an exhibition or contest, though many times quantities of houses are made and sold for other purposes, such as raising money for athletic suits for the school teams.

At Cleveland, Ohio, a large number of houses such as are shown in Figs. 15-19 were made for the city Bird Lovers' Association to be placed in the city parks. The boys received the profits of the sale after materials were paid for. In the Mercer Center, Seattle, Wash., the boys wanted suits for the "team." Bird houses were made in dozen lots for a large department store, and soon the boys had all the money the suits cost. Fig. 36 shows a group of 7th grade boys with the houses made in two class periods of two hours each. At St. Paul, Minn., the annual exhibit has become a larger affair than the automobile show. This year it will be held in the city auditorium which seats 10,000 people. The city council will pay the rent of this building for a week and the boys will see that it is filled with bird houses. Up to date (March 11, 1916) over $1,000 worth of orders have been taken for houses to be delivered after the exhibition. Fig. 62 shows the palm room at the St. Paul exhibit in 1915. The county making the most bird houses in 1915, so far as has been reported, was Allegheny County, Pa., where approximately 15,000 houses were produced. Fig. 67 shows the prize winners in a department store contest at Pittsburgh, Pa., while an exhibit in the same city is shown in Fig. 66.

[Pg 55]

FIG. 66. THE PITTSBURGH EXHIBIT.

FIG. 67. PRIZEWINNERS IN DEPARTMENT STORE CONTEST.

Space will not permit giving extended rules for such contests since the rules must vary with each city. Briefly, there should be provision made to give all competitors an equal chance. Boys of the 6th grade should meet others of that grade. Prizes may be awarded for the best houses made for the more common birds, such as wrens, bluebirds, and martins. These [Pg 57] should be judged as to adaptability or fitness to purpose, amount of protection afforded to birds, good workmanship and artistic merit. A prize might be awarded to the boy whose house has the first pair of birds nesting in it. Prizes may be of many kinds, but tools and books, as well as cash prizes are often given by local business men.

WHERE MORE INFORMATION MAY BE OBTAINED.

U. S. Dept. of Agriculture, Division of Publications: Bird Houses and How to Build Them, Bulletin No. 609; Fifty Common Birds, Bulletin No. 513 (15 cents); The English Sparrow as a Pest, Bulletin No. 493.

Magazines which have published articles on birds and bird houses: Bird Lore; Country Life; The Craftsman; Elementary School Teacher; Ladies' Home Journal; Manual Training and Vocational Education; Outing; Outlook; School Arts Magazine; Something To Do; The Farm Journal; The National Geographic Magazine; Youths Companion.

National Association of Audubon Societies: Leaflets, photographs, advice.

Liberty Bell Bird Club of The Farm Journal, Philadelphia, Pa.: Leaflets, books, pictures, supplies, inspiration.

Public Library: Reed, "Bird Guide"; Blackburn, "Problems in Farm Woodwork"; Chapman, "Bird Life"; Hiesemann, "How to Attract Wild Birds"; National Geographic Society, "Common Birds of Town and Country"; Trafton, "Methods of Attracting Birds."

Catalogs of Bird House Companies: Audubon Bird House Co., Meriden, New Hampshire; "Bird Architecture" Crescent Co., Toms River, N. J. (20 cents); Joseph H. Dodson, 728 Security Bldg., Chicago, Ill.; "Bird Houses Large and Small," Mathews Mfg. Co., Cleveland, Ohio; Charles E. White, Box 45B, Kenilworth, Ill.; The Wheatley Pottery, 2429 Reading Road, Cincinnati, Ohio.

[Pg 59]

MANUAL TRAINING TOYS

FOR

THE BOY'S WORKSHOP

By HARRIS W. MOORE
Supervisor of Manual Training, Watertown, Mass.

A COLLECTION OF FORTY-TWO PROJECTS OVERFLOWING WITH "BOY" INTEREST

A popular boys' book that is truly educational. The projects are all new in the manual training shop. The text gives instructions for making each project and treats of tools and tool processes. The following is a partial list —

Windmills	Pile Driver	Guns
Kites	Kite String Reel	Whistles
Water Wheels	Cannon	Bow and Arrows
Water Motors	Darts	Swords
Pumps	Buzzers	Boxes
Boat	Tops	Telephone

THIRTY-FIVE FULL-PAGE PLATES OF WORKING DRAWINGS

PRICE, POSTPAID, $1.15

KITECRAFT

AND KITE TOURNAMENTS

By CHARLES M. MILLER
>Assistant Supervisor of Manual Training
>Los Angeles, California

KITES — AEROPLANES — TOURNAMENTS

A comprehensive and reliable treatment of kites and kite flying. Mr. Miller, the author of the book, for a number of years past, has made a wonderful success of kite flying in the schools of Los Angeles, California. The book deals with general kite construction, tells how to make all kinds of kites and how to fly them. Describes kite accessories and how to decorate kites. It also describes the construction and use of moving devices, messengers, suspended figures and appliances, balloons and parachutes.

Four chapters treat of aeroplanes, gliders and model aeroplanes, together with one on propellers, motors, gears and winding devices.

The book contains over 267 illustrations, photographs, drawings, and diagrams.

>PRICE, POSTPAID, $1.50

THE MANUAL ARTS PRESS, PEORIA, ILLINOIS

[Pg 62]

CHOICE BOOKS FOR BOYS

FOR HOME AND SCHOOL LIBRARIES

DESIGN AND CONSTRUCTION IN WOOD.

BY WILLIAM NOYES. A book full of charm and distinction and the first to give due consideration to the esthetic side of woodworking. It is intended to give to beginners practice in designing simple projects in wood and an opportunity to acquire skill in handling tools. The book illustrates a series of projects and gives suggestions for other similar projects together with information regarding tools and processes for making. A pleasing volume abundantly and beautifully illustrated. Price, $1.75.

HANDWORK IN WOOD.

BY WILLIAM NOYES. A handbook for teachers and a textbook for normal school and college students. The best reference book available for teachers of woodworking. A comprehensive and scholarly treatise, covering logging, saw-milling, seasoning and measuring, hand tools, wood fastenings, equipment and care of the shop, the common joints, types of wood structures, principles of joinery, and wood finishing. Price, $2.25.

WOOD AND FOREST.

BY WILLIAM NOYES. Treats of wood, distribution of American forests, life of the forest, enemies of the forest, destruction, conservation and uses of the forest, with a key to the common woods by Filibert Roth. Describes 67 principal species of wood with maps of the habitat, leaf drawings, life size photographs and microphotographs of sections. Contains a general bibliography of books and articles on wood and forest. Profusely illustrated with photographs from the United States forest service and with pen and ink drawings by Anna Gausmann Noyes and photographs by the author. Price, $3.50.

ESSENTIALS OF WOODWORKING.

BY IRA S. GRIFFITH. A clear and comprehensive treatment of woodworking tools, materials, and processes. It is illustrated with photographs and numerous pen drawings. The standard textbook for students beginning woodworking. Price, 75 cents.

ART METALWORK.

BY ARTHUR F. PAYNE. A textbook written by an expert craftsman and experienced teacher. It treats of the various materials and their production, ores, alloys, commercial forms, etc.; of tools and equipments suitable for the work; the inexpensive equipment of the practical craftsman; and of the correlation of art metalwork with design and other school subjects. It describes in detail all the processes involved in making articles ranging from a watch fob to a silver loving-cup. It is abundantly and beautifully illustrated, showing work done by students under ordinary school conditions in a manual training shop. The standard book on the subject. Price, $2.00.

THE MANUAL ARTS PRESS, PEORIA, ILLINOIS

[Pg 63]

CHOICE BOOKS FOR BOYS

THE "PROBLEMS SERIES" FOR EITHER HOME OR SCHOOL USE

PROBLEMS IN FARM WOODWORK.

BY SAMUEL A. BLACKBURN. A book of working drawings of 100 practical problems relating to agriculture and farm life. There are 60 full-page plates of working drawings, each accompanied by a page or more of text treating of "Purpose," "Material," "Bill of Stock," "Tools," "Directions" and "Assembly." A wonderfully practical book. Price, $1.25.

PROJECTS FOR BEGINNING WOODWORK AND MECHANICAL DRAWING.

BY IRA S. GRIFFITH. Consists of working drawings and working directions. The projects are such as have proved of exceptional service where woodworking and mechanical drawing are taught in a thoro, systematic manner in the seventh and eighth grades. The 50 projects in the book have been selected and organized with the constant aim of securing the highest educational results. Price, $1.00.

PROBLEMS IN WOODWORKING.

BY M. W. MURRAY. A convenient collection of good problems consisting of 40 plates of working drawings of problems in bench work that have been successfully worked out by boys in grades seven to nine, inclusive. Price 75 cents.

PROBLEMS IN FURNITURE MAKING.

BY FRED D. CRAWSHAW. Contains 43 full-page working drawings of articles of furniture. Every piece shown is appropriate and serviceable in the home. In addition to the working drawings, there is a perspective sketch of each article completed. There 36 pages of text giving notes on the construction of each project, chapters on the "Design," and "Construction" of furniture, and one on "Finishes." Price, $1.00.

FURNITURE MAKING — ADVANCED PROJECTS IN WOODWORK.

BY IRA S. GRIFFITH. Consists of 50 plates of problems and accompanying notes. It is essentially a collection of problems in furniture making, selected and designed with reference to school use. On the plate with each working drawing is a good perspective sketch of the completed object. In draftsmanship and refinement of design these problems are of superior quality. It is in every respect an excellent collection. Price, 95 cents.

PROBLEMS IN WOOD-TURNING.

BY FRED D. CRAWSHAW. A textbook on the science and art of wood-turning. Contains 25 full-page plates of working drawings covering spindle, face-plate, and chuck turning. It is a clear, practical and suggestive book on wood-turning, and a valuable textbook for students' use. Price, 50 cents.

PROBLEMS IN MECHANICAL DRAWING.

BY CHARLES A. BENNETT. With drawings made by Fred D. Crawshaw. A students' textbook consisting of 80 plates classified into groups according to principle and arranged according to difficulty of solution. Price, 75 cents.

THE MANUAL ARTS PRESS, PEORIA, ILLINOIS

www.ingramcontent.com/pod-product-compliance
Lightning Source LLC
Chambersburg PA
CBHW031446210526
45464CB00005B/2342